ACHIEVE

Key Stage 1 Mathematics

Revision and Practice Questions

Catherine Casey

RISING★STARS

Acknowledgements

Rising Stars would like to thank the following trial schools for their feedback on the Achieve Key Stage 1 series:
Edgewood Primary School, Notts, Henwick Primary School, London, Tennyson Road Primary, Beds, Christ Church
Church of England (VC) Primary School, Wiltshire, Churchfields, The Village School, Wiltshire and Chacewater
Community Primary School, Cornwall.

Photo credits: page 21 © CathyDoi/iStockphoto; page 37 © thumb/iStockphoto.

Every effort has been made to trace all copyright holders, but if any have been inadvertently overlooked, the
Publishers will be pleased to make the necessary arrangements at the first opportunity.

Although every effort has been made to ensure that website addresses are correct at time of going to press, Rising
Stars cannot be held responsible for the content of any website mentioned in this book. It is sometimes possible
to find a relocated web page by typing in the address of the home page for a website in the URL window of your
browser.

ISBN: 978-1-78339-538-5

Text, design, illustrations and layout © 2020 Hodder & Stoughton Limited
First published in 2015 by Hodder & Stoughton Limited
(for its Rising Stars imprint, part of the Hodder Education Group),
An Hachette UK Company
Carmelite House
50 Victoria Embankment
London EC4Y 0DZ

www.risingstars-uk.com

Impression number 10 9

Year 2024 2023

The right of Catherine Casey to be identified as the author of this work had been asserted by her in accordance
with the Copyright, Design and Patents Act 1998.

Author: Catherine Casey
Series Editor: Naomi Norman
Educational consultant: Sarah-Anne Fernandes
Educational advisor: Cherri Moseley
Accessibility reviewer: Vivien Kilburn
Publishers: Kate Jamieson and Gillian Lindsey
Project Manager: Vanessa Handscombe
Editorial: Kim Vernon, Alison Walters and Sarah Bishop
Illustrations: Adam Linley/Beehive Illustration and Judy Brown/Beehive Illustration

Cover design: Burville-Riley Partnership
Text design and typeset by Out of House Publishing
Printed in Great Britain by Ashford Colour Press Ltd

A catalogue record for this title is available from the British Library.

Contents

Introduction

Hello. I'm Seren. Welcome to the Achieve Mathematics Revision and Practice Book.

I'll be your guide as you work through the book. I'll tell you about each topic and what you need to know. Together, we'll revise all the Maths topics that you learned in Years 1 and 2.

We'll also practise some questions for each topic. The 'Team Achieve' will show you how to answer the questions.

Let's meet the team.

Hi, I'm Rhys.

Hello, I'm Leena.

Hello,
I'm Kofi.

Hi,
I'm Zofia.

Let's go.

5

How to use this book

In this section Seren tells you what you will need to know for each topic for the end of Key Stage 1 National Tests.

This question lets you practise the skills for the topic. It is just like a real test question so gives you practice for the end of Key Stage 1 National Tests.

The flow chart shows you how to answer the Let's try question. You can learn the method and use it to answer similar questions in this book and in the National Tests.

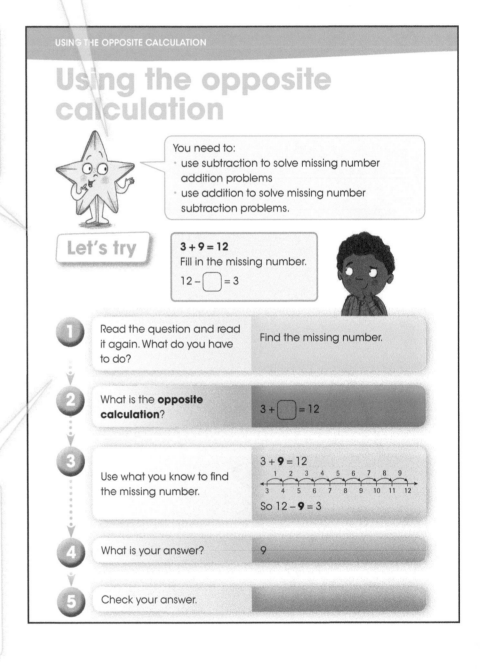

USING THE OPPOSITE CALCULATION

Using the opposite calculation

You need to:
- use subtraction to solve missing number addition problems
- use addition to solve missing number subtraction problems.

Let's try

3 + 9 = 12
Fill in the missing number.
12 – ☐ = 3

1. Read the question and read it again. What do you have to do? — Find the missing number.

2. What is the **opposite calculation**? — 3 + ☐ = 12

3. Use what you know to find the missing number. — 3 + **9** = 12
So 12 – **9** = 3

4. What is your answer? — 9

5. Check your answer.

Addition, subtraction, multiplication and division

Your turn

1 Use the **3** numbers on the triangle. Write **2** addition and **2** subtraction facts. The first one is done for you.

4

8   12

$4 + 8 = 12$ ☐ − ☐ = ☐

☐ + ☐ = ☐   ☐ − ☐ = ☐

In this section you will find practice questions for you to answer. Try to use the method from the flow chart or the Top tips to help you.

2 **16 + 4 = 20**
Use this fact to fill in the missing numbers.

☐ − 4 = 16

20 − 16 = ☐

3 Fill in the missing numbers.

18 − ☐ = 11

☐ + 7 = 18

4 There were **16** grapes in a bowl. Zofia ate some grapes. There were **4** grapes left. How many grapes did Zofia eat?

a $16 - ☐ = 4$

b Use addition to check your answer.

☐ + ☐ = ☐

These tips give you good ideas on how to answer the questions, or useful information on the topic.

Top tips

• Addition can be done in any order.

• Addition and subtraction are the opposite of each other.

Useful Maths resources to help with your revision can be found online at My Rising Stars.

Counting in multiples

You need to:
- count in multiples of 2, 5 and 10 to 100, forwards and backwards
- count forwards in multiples of 3 to 30
- count in steps of 10 to 100, forward and backward (e.g. 97, 87, 77, 67 ...)

Let's try

Fill in the missing numbers.
0, 3, 6, ☐, ☐, 15

1 Read the question and read it again. What do you have to do?

Count in steps.

2 What steps are you counting in?

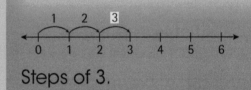

Steps of 3.

3
- Put **6** in your head. Count on **3**.
- Put **9** in your head. Count on **3**.

4 What is your answer?

0, 3, 6, **9, 12,** 15

5 Check your answer.

Your turn

Fill in the missing numbers.

1 0, 2, 4, 6, ☐, ☐, 12

2 45, 40, 35, 30, ☐, 20, ☐, 10

3 12, 15, ☐, ☐, 24, 27, 30

4 14, 24, 34, 44, 54, ☐, 74, ☐, 94

5 98, 88, 78, ☐, ☐, 48, 38

Top tips

- When you count in steps of 10 only the tens digit changes. The ones digit stays the same. For example, 12, 22, 32…
- Use a hundred square to help you when counting in steps.
- Put the number in your head and count the steps on your fingers.

Reading and writing numbers

You need to:
* read and write numbers to 100 in numerals and words.

Let's try

Tick the correct way to write **56** in words.
Tick **one**

five six	☐	sixty-five	☐
fifty-six	☐	fifteen	☐

1 Read the question and read it again. What do you have to do?

Find 56 in words.

2 Say the number.

$56 = 50 + 6$

3 How do I spell **50**?
How do I spell **6**?

fifty
six

4 What is your answer?

fifty-six ☑

5 Check your answer.

Your turn

1 Draw lines to match the numbers to the words. One has been done for you.

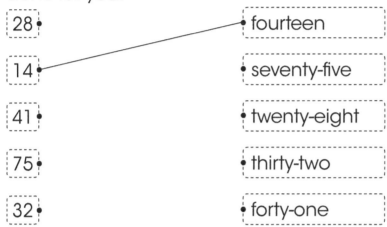

28	fourteen
14	seventy-five
41	twenty-eight
75	thirty-two
32	forty-one

2 Write these numbers in words.
One has been done for you.

89	
13	
57	
16	sixteen

3 Write these words as numbers.
One has been done for you.

twenty-four	
eighteen	
sixty-five	
ninety-one	91

Top tip

If you can't remember how to spell a word, just have a go.
Use your phonics to help you.

Ordering numbers

You need to:
- know the value of each digit in a 2-digit number
- order numbers up to 100
- use the signs <, > and =.

Let's try

Look at these signs:

< > =

Write the correct sign in the box.

87 ☐ 78

1 Read the question and read it again. What do you have to do?

Use the signs to compare the numbers.

2 What do the signs mean?

> is greater than
= equals
< is less than

3 ☀ Is **87** greater than, less than or equal to **78**?

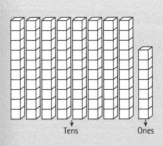

Tens Ones   Tens Ones

☀ How many tens? How many ones?

87 is greater than 78
87 > 78

4 What is your answer? Check your answer.

87 > 78

Your turn

1

Look at the number: 43

How many **tens** are there? How many **ones** are there?

☐ tens   ☐ ones

2

Order these numbers starting with the smallest. One has been done for you.

42   ~~12~~   14   21   41

[12] ☐ ☐ ☐ ☐

smallest ⟶ biggest

3

Look at these signs.

[<] [>] [=]

Write the correct sign in each box.

15 ☐ 51   76 ☐ 67   35 ☐ 35

Top tips

- Remember:
  - = equals
  - < is less than
  - > is greater than
- Remember: the crocodile always eats the bigger number.

23   32

65   56

Number problems

You need to:
- use place value and number facts to solve problems.

Let's try $30 + 70 = \boxed{}$

 1 Read the question and read it again. What do you have to do?

Add together 30 and 70.

 2 What **number fact** can you use?

$3 + 7 = 10$

3 Use the number fact to solve the calculation.

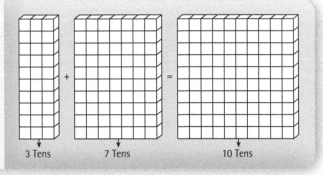

3 Tens   +   7 Tens   =   10 Tens

 4 What is your answer?

100

 5 Check your answer.

Your turn

1
20 + 80 = ☐

2
30 + 30 = ☐

3
100 − 30 = ☐

4
50 − 20 = ☐

5
17 + 3 = ☐

6
20 − 12 = ☐

7
The farmer delivered a box of **70** pears
to the shop and another box with **30** pears.
How many pears were there altogether? ☐

8
Zofia had **80** books on her shelf. She gave **10** to the school book
sale. How many books did she have left? ☐

Top tip

Think: which number fact will help you?

Addition

You need to:
- add 2-digit numbers using the column method.

 **Let's try**

$$45 + 26 = \boxed{}$$

 1
Read the question and read it again. What do you have to do?

Add together 45 and 26.

2
Write the numbers in columns.
Add the **ones**.

	4	5
+	2	6
		1

$5 + 6 = 11$

1 ten   1 one

1 ←The ten from 11 goes here

 3
Add the **tens**.

	4	5
+	2	6
	7	1

$4 + 2 + 1 = 7$
$40 + 20 + 10 = 70$

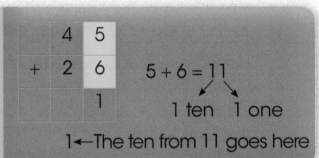

1

Don't forget to add the extra 10.

 4
What is your answer?
Check your answer.

71

Your turn

1

34 + 21 =

	3	4
+	2	1

2

27 + 52 =

	2	7
+	5	2

3

47 + 31 =

	4	7
+	3	1

4

18 + 34 =

	1	8
+	3	4

5

57 + 23 =

	5	7
+	2	3

6

68 + 26 =

	6	8
+	2	6

Top tips

- Only put one digit in each square.
- Line up the tens and ones.

Subtraction

You need to:
* subtract 2-digit numbers using the column method.

$45 - 26 = \boxed{}$

1   Read the question and read it again. What do you have to do?

Take away 26 from 45.

2   Write the numbers in columns.
Subtract the **ones**.

$$\begin{array}{r} {}^3\!\!\not{4} \quad {}^1\!5 \\ - \quad 2 \quad 6 \\ \hline 9 \end{array}$$

We can't take 6 away from 5.

So we need to take a ten from the tens column.

$15 - 6 = 9$

3   Subtract the **tens**.

$$\begin{array}{r} {}^3\!\!\not{4} \quad {}^1\!5 \\ - \quad 2 \quad 6 \\ \hline 1 \quad 9 \end{array}$$

$3 - 2 = 1$
$30 - 20 = 10$

4   What is your answer?
Check your answer.

19

Your turn

1

45 − 21 =

	4	5
−	2	1

2

87 − 16 =

	8	7
−	1	6

3

69 − 34 =

	6	9
−	3	4

4

84 − 17 =

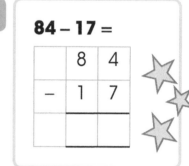

	8	4
−	1	7

5

53 − 37 =

	5	3
−	3	7

6

63 − 26 =

	6	3
−	2	6

Top tips

- Take the smaller number away from the bigger number.
- Line up the tens and ones.

Adding 3 numbers

You need to:
* add 3 one-digit numbers
* use doubles facts and number facts to 20.

$3 + 4 + 3 = \boxed{}$

1 Read the question and read it again. What do you have to do?

Add 3 numbers
$3 + 4 + 3 =$

2 What **number fact** or **doubles fact** can you use?

$3 + 4 + 3 =$
$3 + 3 = 6$

3 What do you have left to add?

$6 + 4 =$

4 What is your answer?

10

5 Check your answer.

Your turn

1 $2 + 6 + 2 = \boxed{}$

2 $3 + 7 + 2 = \boxed{}$

3 $7 + 5 + 7 = \boxed{}$

4 $5 + 9 + 5 = \boxed{}$

5 Leena has **nine** marbles, Zofia has **nine** marbles and Rhys has **seven** marbles. How many marbles do they have altogether? $\boxed{}$

6 Kofi saw **17** blackbirds, **2** robins and **3** starlings in his garden. How many birds did he see altogether? $\boxed{}$

7 $7 + 14 + 6 = \boxed{}$

8 $15 + 5 + 8 = \boxed{}$

Top tips

- Look for pairs that add up to 10, 20 or doubles.
- Draw a circle around any doubles or number facts to help you spot them.
- Remember: addition can be done in any order.

Using the opposite calculation

You need to:
- use subtraction to solve missing number addition problems
- use addition to solve missing number subtraction problems.

Let's try

3 + 9 = 12
Fill in the missing number.
12 − ☐ = 3

 1 Read the question and read it again. What do you have to do?

Find the missing number.

 2 What is the **opposite calculation**?

3 + ☐ = 12

3 Use what you know to find the missing number.

3 + **9** = 12

1 2 3 4 5 6 7 8 9
3 4 5 6 7 8 9 10 11 12

So 12 − **9** = 3

 4 What is your answer?

9

 5 Check your answer.

Your turn

1

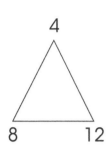

4

8   12

Use the **3** numbers on the triangle. Write **2** addition and **2** subtraction facts. The first one is done for you.

⬚ 4 ⬚ + ⬚ 8 ⬚ = ⬚ 12 ⬚   ⬚ ⬚ − ⬚ ⬚ = ⬚ ⬚

⬚ ⬚ + ⬚ ⬚ = ⬚ ⬚   ⬚ ⬚ − ⬚ ⬚ = ⬚ ⬚

2

16 + 4 = 20

Use this fact to fill in the missing numbers.

⬚ − 4 = 16

20 − 16 = ⬚

3

Fill in the missing numbers.

18 − ⬚ = 11

⬚ + 7 = 18

4

There were **16** grapes in a bowl. Zofia ate some grapes. There were **4** grapes left. How many grapes did Zofia eat?

a 16 − ⬚ = 4

b Use addition to check your answer.

⬚ + ⬚ = ⬚

Top tips

- Addition can be done in any order.

- Addition and subtraction are the opposite of each other.

Addition and subtraction problems

You need to:
- solve 2-step problems involving addition and subtraction.

Let's try

There were **33** 'Well done' stickers.
Kofi got **6**. Zofia got **9**.
How many stickers were left?

1 | Read the question and read it again. What do you have to do? | Take away 6 and 9 from 33.

2
- How many stickers were there to start with?

  33

- How many stickers were given out altogether?

  $6 + 9 = 15$

- Write the number sentence.

  $33 - 15 =$

3
- Put the numbers into columns with the bigger number on top. Subtract the **ones**.

  You can't take 5 from 3. So we need to take a ten from the tens column. $13 - 5 = 8$

	²3̶	¹3
−	1	5
		8

- Subtract the **tens**.

  $20 - 10 = 10$
  $2 - 1 = 1$

	²3̶	¹3
−	1	5
	1	8

What is your answer?
Check your answer.

18

24

Your turn

1

There are **34** strawberries. Rhys picks **17** strawberries. Kofi picks **5** strawberries. How many strawberries are left?

17 + 5 = ⬚

34 − ⬚ = ⬚

2

35 children are in a swimming pool. **8** have blue goggles. **5** have pink goggles. How many children don't have goggles?

8 + 5 = ⬚

35 − ⬚ = ⬚

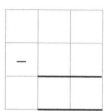

3

The bus has **59** seats. **5** boys sit in the bus. **12** girls sit in the bus. How many seats are left?

5 + 12 = ⬚

59 − ⬚ = ⬚

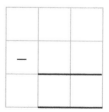

Top tip

Put the bigger number on the top for a subtraction problem.

Multiplication

You need to:
- know odd and even numbers
- recall and use multiplication facts for the 10, 2 and 5 multiplication tables and use the appropriate signs (× and =).

Let's try

$5 \times 3 = \boxed{}$

1 Read the question and read it again. What do you have to do?

Find 5 lots of 3.

2 Draw an **array**.

3 Let's count in **threes**.

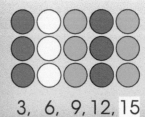

3, 6, 9, 12, 15

4 What is your answer?

15

5 Check your answer.

Your turn

1

Circle the **even** numbers.

17   32   84   51   66

2

5 × 6 = ☐

Draw an array to help you.

3

Match the number sentences. One has been done for you.

2 × 3 =	5 + 5 + 5 + 5 + 5 + 5 =
5 × 6 =	10 + 10 =
10 × 2 =	2 + 2 + 2 =

4

Match the number sentences. One has been done for you.

3 × 2 =	10 × 8 =
7 × 2 =	2 × 3 =
6 × 5 =	2 × 7 =
8 × 10 =	5 × 6 =

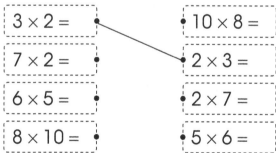

Top tips

- Multiplication can be done in any order: 2 × 3 = 3 × 2.
- Try to learn your 2, 3, 5 and 10 times tables.
- An even number can be divided equally by 2. An odd number can not be divided equally by 2.

Division

You need to:
* recall division facts for the 10, 2 and 5 multiplication tables and use the appropriate signs (÷ and =).

Let's try $12 \div 2 = \boxed{}$

1 Read the question and read it again. What do you have to do?

Share 12 into groups of 2.

2 Draw a picture.

3 How many groups are there?

4 What is your answer?

6

5 Check your answer.

Your turn

1

$8 \div 2 = \boxed{}$

2

$10 \div 5 = \boxed{}$

3

$20 \div 10 = \boxed{}$

4

$16 \div 2 = \boxed{}$

5

15 children get into teams of **5**. How many teams are there? $\boxed{}$

6

Match the number sentences. One has been done for you.

$10 \div 2 = 5$ $8 \times 2 = 16$

$16 \div 2 = 8$ $6 \times 5 = 30$

$40 \div 10 = 4$ $5 \times 2 = 10$

$30 \div 5 = 6$ $4 \times 10 = 40$

Top tip

Try to learn your 2, 5 and 10 times tables and the matching division calculations.

For example, $3 \times 5 = 15$

$15 \div 5 = 3$

$15 \div 3 = 5$

Multiplication and division problems

You need to:
- use the symbols ÷, × and =
- solve multiplication and division problems.

There are **8** children coming to the party. They eat **2** sandwiches each. How many sandwiches do they eat altogether?

1 Read the question and read it again. What do you have to do?

Find out how many sandwiches the children eat.

2 Write a number sentence.

$8 \times 2 =$

3 Draw an **array**.

How many dots?

16

4 What is your answer? Check your answer.

16

Your turn

1

Kofi buys **2** bags of apples. There are **5** apples in each bag.
How many apples does he buy?

[2] [x] [5] = [ ]

2

There are **15** children in the dance class. The teacher makes
3 groups. How many children are there in each group?

[15] [÷] [3] = [ ]

3

Leena joins **5** pieces of train track.
Each piece of train track is **10 cm** long.
How long is the train track?

[ ] [ ] [ ] = [ ] cm

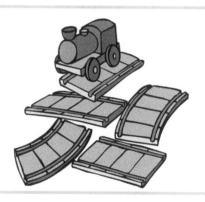

4

Zofia buys **3** bottles of water. Each bottle holds **2 litres**. How many
litres of water does Zofia have?

[ ] [ ] [ ] = [ ] litres

Top tips

- Decide whether it is a multiplication or division problem.
- Write a number sentence to help you.

Fractions of shapes

You need to:
- find, name and write the fractions half $\left(\frac{1}{2}\right)$, quarter $\left(\frac{1}{4}\right)$, third $\left(\frac{1}{3}\right)$, three-quarters $\left(\frac{3}{4}\right)$
- recognise that $\frac{1}{2}$ is equal to $\frac{2}{4}$.

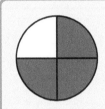 What fraction of the shape is shaded? ☐

Let's try

1 Read the question and read it again. What do you have to do? | Find what fraction of the shape is shaded.

2 How many parts are there? | There are 4 parts. The shape is split into quarters.

3 How many of the parts are shaded? | 3 parts are shaded.

4 What is your answer? | $\frac{3}{4}$

5 Check your answer.

Your turn

1 What fraction is shaded? ☐

2 What fraction is **not** shaded? ☐

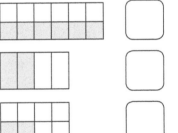

3 Look at the squares. Tick the one with $\frac{3}{4}$ shaded.

☐ ☐ ☐

4 Look at the rectangles. Tick the **two** shapes that have $\frac{1}{2}$ shaded.

5 Colour $\frac{1}{4}$ of the shape.

Top tips

Fractions are **equal** parts of one whole.

$\frac{1}{2}$ = one half

$\frac{1}{4}$ = one quarter

$\frac{1}{3}$ = one third

$\frac{3}{4}$ = three-quarters $\frac{1}{4} + \frac{1}{4} + \frac{1}{4}$

$\frac{1}{2}$ is equal to $\frac{2}{4}$

Fractions of numbers

You need to:
- find, name and write the fractions half $\left(\frac{1}{2}\right)$, quarter$\left(\frac{1}{4}\right)$, third $\left(\frac{1}{3}\right)$, three-quarters $\left(\frac{3}{4}\right)$
- know that $\frac{1}{2}$ is equal to $\frac{2}{4}$.

Let's try

$\frac{1}{2}$ of **12** = ☐

1 Read the question and read it again. What do you have to do?

Find $\frac{1}{2}$ of 12.

2
- What is the whole amount?
- How many **equal** groups do you need to make?
- Draw a **fraction bar** to help you.

The whole amount is 12.

There are two equal groups.

3 Share **12** dots equally between the **2** halves.

Count to 12 as you put one dot in each group. Count the number of dots in each group.

Each group has 6 dots.

4 What is your answer?
Check your answer.

$\frac{1}{2}$ of 12 = 6

Your turn

1

$\frac{1}{2}$ of **8** = ☐

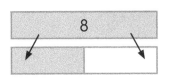

2

$\frac{1}{3}$ of **12** pencils = ☐

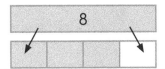

3

Rhys has **16** cars. He gives Leena **one-quarter** of the cars. How many cars does Rhys give Leena? ☐

4

$\frac{3}{4}$ of **8** cakes = ☐

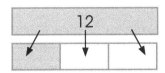

5

Look at these fractions.

$\frac{3}{4}$ $\frac{2}{4}$ $\frac{1}{4}$ $\frac{1}{3}$ $\frac{1}{2}$

Circle the **two** fractions that are **equal**.

Top tip

When sharing into groups, make sure that the groups are always equal.

Length

You need to:
- compare and order lengths
- use a ruler to measure length.

Let's try

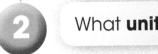

How long is the pencil?

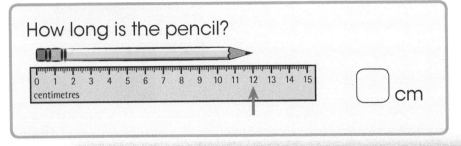 

☐ cm

1 Read the question and read it again. What do you have to do? | Measure the length of the pencil.

2 What **units** are you using? | centimetres (cm)

3 What cm mark is by the point of the pencil?
Make sure the other end of the pencil is lined up with 0 before you start measuring.

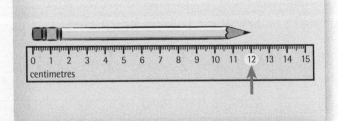

4 What is your answer? | 12 cm

5 Check your answer.

Your turn

1 Put the lengths in **order**, starting with the smallest.

62 m   50 m   14 m   3 m   26 m

☐   ☐   ☐   ☐   ☐

shortest ⎯⎯⎯⎯⎯⎯⎯⎯⎯⎯⎯⎯⎯⎯→ longest

2 How long is this piece of ribbon?

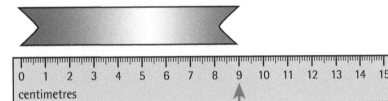

☐ cm

3 Use a ruler to measure the length of the key.

☐ cm

4 A swimming pool is **3 m** wide. It is **twice** as long. How long is the swimming pool? ☐ m

Top tips

- Remember the units in your answer.
- 1 metre = 100 centimetres.

Mass

You need to:
- compare and order mass
- measure how heavy an object is.

Let's try

How heavy is the parcel?

[ ] g

1 Read the question and read it again. What do you have to do?

Find the mass of the parcel.

2 What **units** are you using?

grams (g)

3 Look at the scales. Where is the arrow pointing?

The arrow is pointing at the 4th mark after 100 g.
There are 5 marks between 100 g and 150 g, so each mark is worth 10 g.
10 g + 10 g + 10 g + 10 g = 40 g
100 g + 40 g = 140 g

4 What is your answer? Check your answer.

140 g

Your turn

1 Kofi weighed the toys.

200 g   300 g   5 g   95 g   50 g

a Which toy has the **biggest** mass? ☐

b Which toy has the **smallest** mass? ☐

2 How heavy is the sugar? ☐ g

3 Look at the signs.

<   >   =

Use the signs to complete the boxes.

81 kg ☐ 18 kg

30 g ☐ 30 g

28 g ☐ 38 g

Top tips

- Mass is how heavy something is.
- We measure mass in grams (g) and kilograms (kg).
- Kilo means 1000. This helps to remember that 1000 g = 1 kg.
- Remember to record the units too.

Capacity

You need to:
* measure volume or capacity
* compare and order capacities.

Let's try

How much water is in the jug?

 ml

1 Read the question and read it again. What do you have to do?

Read the scale to measure the volume of the water.

2 What **units** are you using?

millilitres (ml)

3 What **ml** mark is the top of the water by?

The top of the water is halfway between 300 ml and 400 ml. The difference between 400 and 300 is 100

$400 - 300 = 100$ ml

$\frac{1}{2}$ of 100 is 50

$300 + 50 = 350$ ml

4 What is your answer?

350 ml

5 Check your answer.

Your turn

1 How much medicine is in the tube? ⬚ ml

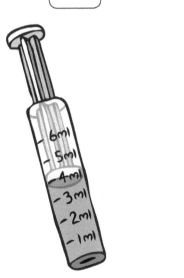

2 How much water is in the measuring cylinder? ⬚ ml

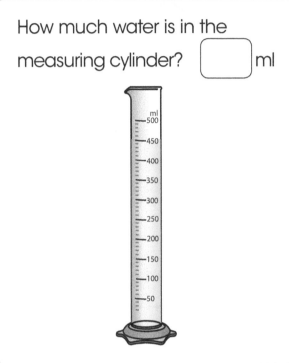

3 **400 ml** of lemonade is in the jug. Leena pours in **300 ml** of orange juice. How much drink is in the jug altogether? ⬚ ml

4 There is **600 ml** of juice in the jug. Rhys drinks **half** the juice. How much juice is left in the jug? ⬚ ml

Top tips

- We measure capacity in litres (l) and millilitres (ml).
- 1 litre = 1000 ml.
- Remember to record the units too.

Temperature

You need to:
- measure how hot or cold something is
- use the units degrees Celsius (°C).

Let's try

Look at the thermometer. What is the temperature?

☐ °C

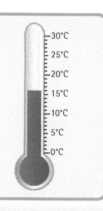

1 Read the question and read it again. What do you have to do?

Read the temperature.

2 What **units** are you using?

degrees Celsius (°C)

3 Look carefully at the scale. What value does each mark have?

There are 5 marks up to the 5°C label so each mark is 1°C.

4 Where is the red line?

1 mark above 15°C
15 + 1 = 16

5 What is your answer? Check your answer.

16°C

Your turn

1 Look at the thermometer. What temperature is shown?

[ ] °C

— 20°C

— 15°C

— 10°C

— 5°C

—0°C

2 Look at the thermometer. What temperature is shown?

[ ] °C

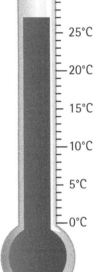

— 30°C

— 25°C

— 20°C

— 15°C

— 10°C

— 5°C

—0°C

3 Look at the thermometer. What temperature is shown?

[ ] °C

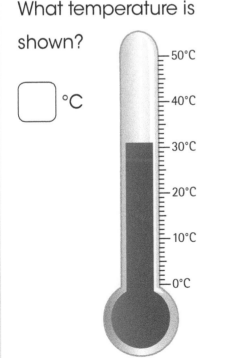

— 50°C

— 40°C

— 30°C

— 20°C

— 10°C

— 0°C

Top tips

- Always check the scale. How much is each mark on it worth?
- Remember: temperature means how hot or cold something is.

43

Money

You need to:
- know the value of different coins
- use the symbols for pounds (£) and pence (p)
- combine amounts to make a particular value and find different combinations of coins to equal the same amounts of money.

Let's try

Which coins could you use to make **36p**?

1 Read the question and read it again. What do you have to do?

Use coins to make 36p.

2 What coins are there?

3 Add coins together to make **36p**.

There are lots of different ways to do this.

 = 36p

4 What is your answer?

20p, 10p, 5p, 1p

5 Check your answer.

Your turn

1

a How much money is in Leena's purse? ▢

b Use different coins to make the same amount.

▢

2

a Zofia has these coins. How much has she got? ▢

b Use different coins to make the same amount.

▢

3

a Which coins could you use to make **74p**? ▢

b Use different coins to make **74p**. ▢

Top tips

- Use a decimal point to separate the pounds and pence, e.g. £3.75.
- Remember: if you use a pound sign, you do not need a pence sign too.
- Remember: £1 = 100p.

Money problems

You need to:
- solve money problems
- solve money problems where change is given.

Let's try

Rhys has **30p**. He buys a drink for **14p** and a cake for **12p**. How much change does Rhys get from **30p**? ⬚

1 Read the question and read it again. What do you have to do?

Find out how much change Rhys gets.

2 How much did Rhys spend?

14p + 12p = 26p

3
- ★ **Subtract** the amount Rhys spent away from 30p.
- ★ What **calculation** do you need to do?

30p – 26p =

We would count from 26 up to 30.

4 Use a **number line** to help you.

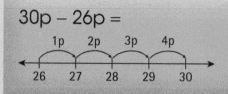

30p – 26p =

5 What is your answer? Check your answer.

4p

Your turn

1 Leena buys a bag of sweets for **21p** and a magazine for **36p**.
How much does she spend altogether? ☐ p

2 Kofi has **30p**. Tick the **two** items he can buy together.

☐ ☐ ☐ ☐

3 Rhys has **70p**. He spends **25p**. How much does he have left?
☐ p

4 Zofia buys a toothbrush for **£3.50** and toothpaste for **£5**. Work out how much change she gets from **£10**.

£3.50 + £5 = ☐

£10 − ☐ = ☐

Top tips

- Write a number sentence to help you.
- Remember: 100p = £1.

Telling the time

You need to:
- read and write times
- draw hands on a clock face.

Let's try

What time is it? [ ]

| 1 | What are you being asked to do? | Read the time on the clock. |

| 2 | Where is the long hand? |  quarter past |

| 3 | Where is the short hand? | 5 |

| 4 | What is your answer? | Quarter-past 5 |

| 5 | Check your answer. | |

Your turn

1 Draw lines to match the correct times to the clocks.

One has been done for you.

half-past 11

quarter-to 5

ten-past 3

quarter-past 8

2 Draw the hands on the clock face to show the correct time.

a 2 o'clock

b half past 8

c twenty-five past 9

3 Kofi gets on the train at **ten-past 4**. He is on the train for **2** hours. Draw the hands on the clock to show the time Kofi gets off the train.

Top tips

- The minute hand is the long hand and the hour hand is the short hand.
- 1 hour = 60 minutes; half an hour = 30 minutes; quarter of an hour = 15 minutes.

Ordering time

You need to:
* order units of time
* know the days of the week and months of the year.

Let's try

Put the units of time in order.

1 month	1 week	1 hour	1 year	1 day
☐	☐	☐	☐	☐

shortest ───────────────────→ longest

1 Read the question and read it again. What do you have to do? → Put the units of time in order.

2 What order do they need to be in? → Starting with the shortest and finishing with the longest.

shortest ──────────→ longest

3 How long is each unit? →
1 hour = 60 minutes
1 day = 24 hours
1 week = 7 days
1 month = about 4 weeks
1 year = 12 months

4
★ Which unit of time is the shortest? → 1 hour
★ Which unit of time is next? → 1 day, 1 week, 1 month
★ Which unit of time is the longest? → 1 year

5 What is your answer? Check your answer. → 1 hour, 1 day, 1 week, 1 month, 1 year

Your turn

1 Put the days of the week in order. One has been done for you.

Wednesday   Monday

Thursday

Saturday

Monday

Friday

Sunday

Tuesday

2 Match the event to the unit of time. One has been done for you.

Event   **Unit of time**

putting on shoes   hours

a plant growing   minutes

reading a book   seconds

a car journey   weeks

Top tips

60 seconds = 1 minute   7 days = 1 week

60 minutes = 1 hour   52 weeks = 1 year

24 hours = 1 day

2-D shapes

You need to:
- name 2-D shapes
- count the number of sides and vertices
- find vertical lines of symmetry
- compare and sort common 2-D shapes and everyday objects.

Let's try

Fill in the blanks.

A hexagon has ⬜ sides and ⬜ vertices.

hexagon

1 Read the question and read it again. What do you have to do?

Find the number of sides and the number of vertices.

2 Count the **sides**.

 6 sides

3 Count the **vertices**.

 6 vertices

4 What is your answer? Check your answer.

A hexagon has 6 sides and 6 vertices.

Your turn

1 Draw a line from the shape to its name.

• triangle

• square

• circle

• hexagon

2 Look at the shapes below. Put a cross on the shape that is **not** a pentagon.

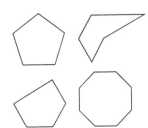

3 Fill in the blanks.

An octagon has ☐ sides and ☐ vertices.

4 Look at the shapes. Put a cross on the shape that does **not** have a line of symmetry.
You can use a mirror to find the lines of symmetry.

Top tips

• *Octo* means 8. Think of an octopus with 8 legs.
• Count the number of sides to identify the shape.

3-D shapes

You need to:
- name 3-D shapes
- count the number of faces, edges and vertices on a 3-D shape
- compare and sort common 3-D shapes
- find the 2-D shapes on the 3-D shapes.

Let's try

How many vertices does a square-based pyramid have? ☐

1 Read the question and read it again. What do you have to do?

Find the number of vertices on the shape.

2 What is a **vertex**?

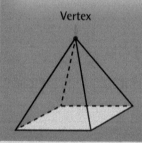

Vertex

3 Count the number of vertices.

There are 5 vertices.

4 What is your answer? Check your answer.

5

Your turn

1 Draw a line to link the shapes to the correct label.

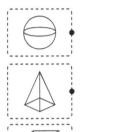

- cuboid
- sphere
- square-based pyramid

2 Look at the pictures.
Put a cross on the one
that is **not** a cone.

3 Fill in the blanks.

Picture	Shape	Faces	Edges	Vertices
	square-based pyramid		8	5
	cuboid	6	12	
	cylinder			0
	triangular-based pyramid		6	4

Top tip

Remember: you cannot see *all* sides of a 3-D shape in a picture.

Repeating patterns

You need to:
- continue a repeating pattern.

Let's try

Draw the next shape in this repeating pattern.

1   Read the question and read it again. What do you have to do?

Find the next shape in the pattern.

2   What is the **pattern** before it starts again?

3   How many circles are in the pattern?

What is the last shape in the diagram?

2 circles.

It is the 1st circle in the pattern, so we need another circle.

4   What is your answer? Check your answer.

Your turn

Draw the next shape in the pattern.

1

2

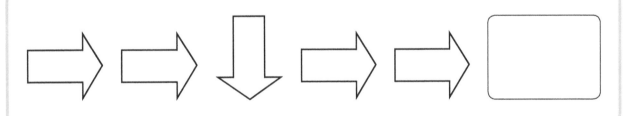

3

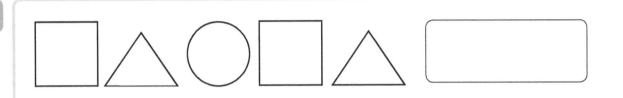

4

5

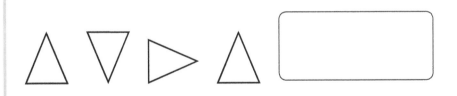

Top tip

When you repeat a pattern, you repeat the objects in the same order.

Directions and movement

You need to:
- describe positions, directions and movement
- know left and right
- recognise movement in a straight line
- understand rotation as a turn, and right angles for quarter turn and half turn.

Let's try

This shape is rotated **clockwise** through a **three-quarter** turn.
What will the shape look like **after** it has been turned?
Tick **one**.

1	Read the question and read it again. What do you have to do?	Imagine the shape rotated clockwise through a quarter turn.

2	✶ What **direction** do you need to turn the shape in? ✶ How far do you need to turn the shape?	clockwise three quarter turns

3	Imagine the shape rotating.	Pick a point on the shape to help you imagine it turning. Count the turns. 

4	What is your answer? Check your answer.	

Your turn

1

This shape is rotated **anti-clockwise** through a quarter turn.
What will the shape look like **after** it has been turned? Tick **one**.

2

This shape is rotated **clockwise** through one right-angle. What
will the shape look like **after** it has been turned? Tick **one**.

3

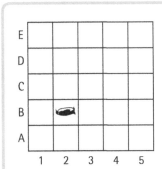

The penguin moves **three** squares **up**,
then **one** square **left**. Tick the square he
lands in.

Top tips

* Use your hands to remember left and right.
* Try turning this book as you follow each
instruction.

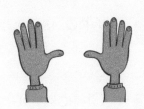

Pictograms and block diagrams

You need to:
- interpret simple pictograms (each symbol is equivalent to one)
- interpret block diagrams (where the scale is divided into ones or multiples of two).

This block diagram shows our favourite colours.

How many **more** children chose blue than red?

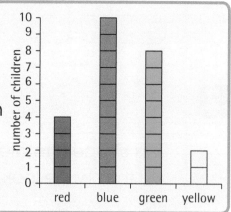

Let's try

1	Read the question and read it again. What do you have to do?	Find out how many more children like blue than red.
2	⋆ How many children like blue? ⋆ How many children like red?	10 4
3	What calculation do you need to do?	10 − 4 = 6
4	What is your answer?	6
5	Check your answer.	

Your turn

1 This pictogram shows children's favourite zoo animals.

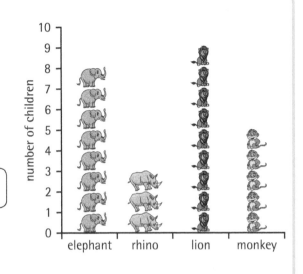

a How many children like a monkey best? ☐

b Which is the **most** popular zoo animal? ☐

c How many **more** children like elephants than rhinos? ☐

2 This block diagram shows our favourite hobbies.

a **5** children like reading. Draw this block on the diagram.

b How many children like computer games? ☐

c Which was the **least** popular hobby? ☐

d How many **more** children like football than reading? ☐

Top tips

- *Least popular* means the group with the smallest amount.
- *Most popular* means the group with the greatest amount.
- Remember to count the objects carefully.

Tally charts and tables

You need to:
- interpret tally charts
- interpret simple tables.

Let's try

This tally chart shows the pets some children have.

Pet	Tally
cat	ⵑⵑ ⵑⵑ ⵑⵑ|
dog	ⵑⵑ |||
fish	||
rabbit	|||

How many **more** children have a dog than have a fish?

1 Read the question and read it again. What do you have to do?

Find out how many more children have a dog than a fish.

2
- How many children have a dog?
- How many children have a fish?

ⵑⵑ |||
5 + 3 = **8**

||
2

3 What calculation do you need to do?

8 − 2 = 6

4 What is your answer? Check your answer.

6

Your turn

1

This tally chart shows our favourite fruit.
6 children liked oranges.
Draw the tally on the chart.

Fruit	Tally
apple	ⅢⅢ ⅢⅢ ⅢⅢ Ⅰ
orange	
banana	ⅢⅢ ⅢⅢ Ⅰ
grapes	ⅠⅠ

2

Use the tally chart to fill in the table.

Our favourite fruit

Fruit	Number of children
apple	
	6
banana	
grapes	

3

How many children like apples best? ☐

4

Which is the **most** popular fruit? ☐

5

Which is the **least** popular fruit? ☐

6

How many **more** children liked apples than bananas? ☐

Remember:
Ⅰ = 1
ⅢⅢ = 5

Answers

Page 9   Counting in multiples
1. 8, 10   2. 25, 15   3. 18, 21
4. 64, 84   5. 68, 58

Page 11   Reading and writing numbers

2. 89 eighty-nine, 13 thirteen, 57 fifty-seven
3. twenty-four 24, eighteen 18, sixty-five 65

Page 13   Ordering numbers
1. 4 tens, 3 ones
2. 14, 21, 41, 42
3. $15 < 51$, $76 > 67$, $35 = 35$

Page 15   Number problems
1. 100   2. 60   3. 70   4. 30
5. 20   6. 8   7. 100   8. 70

Page 17   Addition
1. 55   2. 79   3. 78
4. 52   5. 80   6. 94

Page 19   Subtraction
1. 24   2. 71   3. 35
4. 67   5. 16   6. 37

Page 21   Adding 3 numbers
1. 10   2. 12   3. 19   4. 19
5. 25   6. 22   7. 27   8. 28

Page 23   Using the opposite calculation
1. $8 + 4 = 12$, $12 - 8 = 4$, $12 - 4 = 8$
2. 20, 4   3. 7, 11
4. a 12   b $12 + 4 = 16$

Page 25   Addition and subtraction problems
1. 12   2. 22   3. 42

Page 27   Multiplication
1. 32, 84, 66   2. 30
3. $2 \times 3 =$ $5 + 5 + 5 + 5 + 5 + 5 =$
    $5 \times 6 =$ $10 + 10 =$
    $10 \times 2 =$ $2 + 2 + 2 =$
4. $3 \times 2 =$ $10 \times 8 =$
    $7 \times 2 =$ $2 \times 3 =$
    $6 \times 5 =$ $2 \times 7 =$
    $8 \times 10 =$ $5 \times 6 =$

Page 29   Division
1. 4   2. 2   3. 2
4. 8   5. 3

6. $10 \div 2 = 5$ $8 \times 2 = 16$
    $16 \div 2 = 8$ $6 \times 5 = 30$
    $40 \div 10 = 4$ $5 \times 2 = 10$
    $30 \div 5 = 6$ $4 \times 10 = 40$

Page 31   Multiplication and division problems
1. 10   2. 5
3. $5 \times 10 = 50\,cm$ 4. $3 \times 2 = 6\,litres$

Page 33   Fractions of shapes
1. $\frac{1}{3}$ 2. $\frac{1}{4}$ 3.
4. 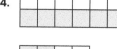   5.

Page 35   Fractions of numbers
1. 4   2. 4   3. 4
4. 6   5. $\frac{2}{4}$ $\frac{1}{2}$

Page 37   Length
1. 3 m, 14 m, 26 m, 50 m, 62 m
2. 9 cm
3. 6 cm   4. 6 m

Page 39   Mass
1. a Dolly (300 g)   b Lego brick (5 g)
2. 180 g
3. $81\,kg > 18\,kg$
    $30\,g = 30\,g$
    $28\,g < 38\,g$

Page 41   Capacity
1. 4 ml   2. 450 ml
3. 700 ml   4. 300 ml

Page 43   Temperature
1. 8°C   2. 27°C   3. 31°C

Page 45   Money
1. a 33p
    b any combination of coins that make 33p
      e.g.: 20p + 5p + 5p + 2p + 1p
2. a £3.15
    b any combination of coins that make
      £3.15 e.g.: £1 + £1 + £1 + 10p + 5p
3. a any combination of coins that make 74p
      e.g.: 50p + 20p + 2p + 2p
    b any other combination e.g.: 50p + 10p + 10p +
      2p + 2p

Page 47   Money problems
1. 57p   2. Drink and cake
3. 45p   4. £1.50